Smoke and Roses

A STEAMPUNK LANGUAGE OF FLOWERS

Olivia Wylie

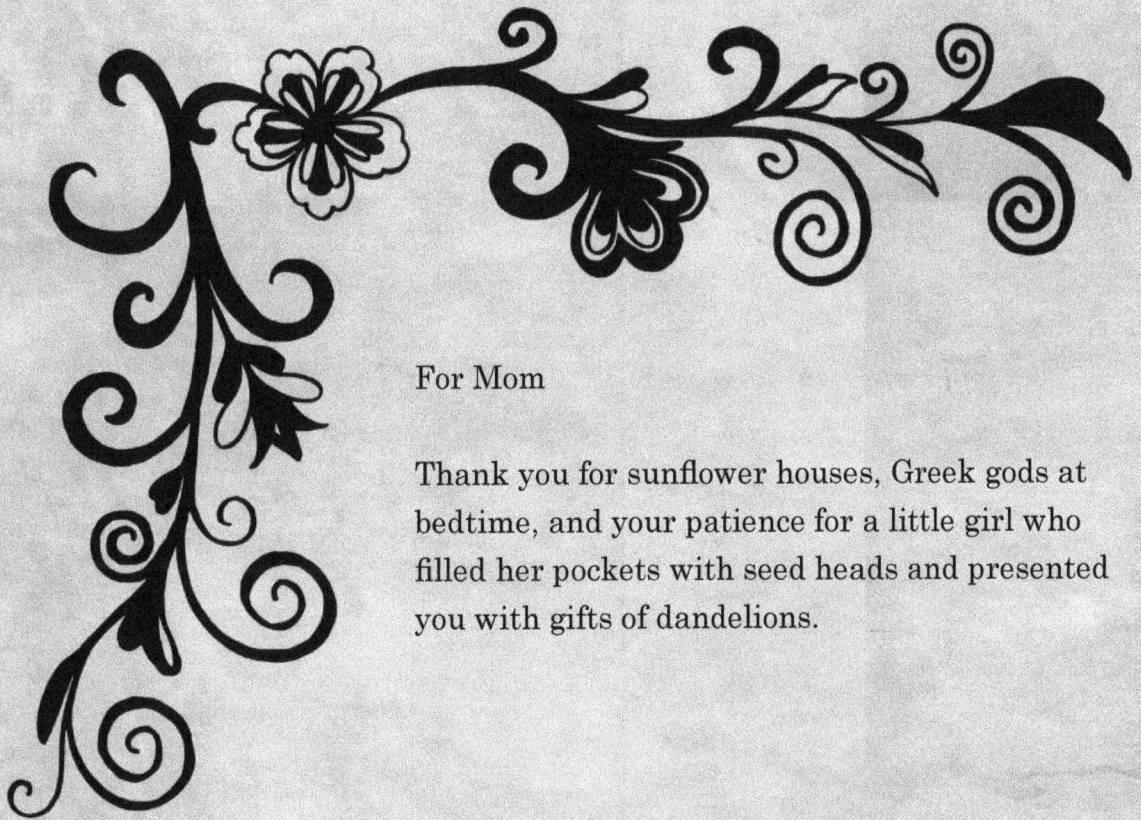

For Mom

Thank you for sunflower houses, Greek gods at bedtime, and your patience for a little girl who filled her pockets with seed heads and presented you with gifts of dandelions.

Contents

Introduction

Our highest assurance of the goodness of Providence seems to me to rest in the flowers. All other things, our powers, our desires, our food, are all really necessary for our existence in the first instance. But this rose is an extra . Its smell and its color are an embellishment of life, not a condition of it. It is only goodness which gives extras, and so I say again that we have much to hope from the flowers.

—Sir Arthur Conan Doyle, The Naval Treaty, 1894

Flowers have been growing with us since human civilization began. The names we give them project our own thoughts and fears upon them: Bachelor's Button, Feverfew, Love Lies Bleeding, Forget Me Not.

Every flower tells a tale. Some whisper of scandal, some sigh of broken hearts and some murmur of love yet to come. Flowers embellish our lives, adding grace notes of color and scent. Perhaps the more we feel life constrain us, the further we seek to embellish it.

That may be why a set of associations between blooms and human motivations grew out of the writings of the Romantic poets to flower during the reign of Queen Victoria. What we call the Language of Flowers today took root from a seed planted by Lady Mary Wortley Montagu in *The Turkish Embassy Letters,* penned in 1763. In these letters, she wrote of the Turkish custom of *selam,* which she described as a symbolic code making it possible for sweethearts to send messages 'without ever inking your fingers'. She was mostly wrong in her surmises, but the idea sprouted in European soil all the same. In France, a lady using the pen name Charlotte de Latour published *Le Langage des Fleurs* at the end of 1819, and a craze took root.

For the aspirational classes, the idea of combining poetry and the new luxury of exotic blooms brought from far off places to be grown in modern hot houses was deeply appealing. This was the Age of Botany as well as the Age of Industry: Carl Linnaeus had finished creating the binomial nomenclature system that could accurately describe every plant known in 1753. Technology was improving, and it was possible with the use of glass houses and Wardian Cases to keep rare blooms from around the world alive despite European weather.

Drugs were pouring into European pharmacopoeias. Every novel bloom was a sign that the world was full of wonders yet to be discovered. Versions of the Language of Flowers soon included exotic beauties like Amaryllis and Orchid, creating a touch of class, a dash of intrigue and the faintest whiff of scandal.

The Victorians ate it up.

Soon, lists of flower meanings began circulating in salons and were passed around in parlors, especially in London. The trend was particularly exciting for women, as the study and love of flowers was viewed by society as a 'noble pursuit', appropriate for ladies. Women with an interest in the sciences latched onto botany in droves, and ladies of a literary persuasion began to use flowers to symbolize underlying circumstances in their works.

Flowers also allowed people in this rigid society to discuss the risque in safety, through poetry. What young people dared not say aloud, they conveyed in bouquets.

In this illustrated volume, I have collected just a short list of flowers and their meanings, doing my best to create an alphabet of flowers that will suit situations encountered in daily life. There are many ways in which our modern day mirrors that time more than a century ago. In an era of rapid industrialization and dangerously mismanaged cities growing ever wider and dirtier, blossoms and their imagery shine like beacons in the grime. People in the Age of Victoria looked for reasons to stop and smell the flowers if they could.

Perhaps we can do the same.

AMARYLLIS

Hippeastrum

Standing tall above other blooms, amaryllis holds its head high. Well it might; amaryllis comes from the Greek word "amarysso," which means "to be splendid", and it seems to revel in the appellation. The German botanist Eduard Friedrich Poeppig discovered the amaryllis hippeastrum growing on a Chilean hillside in the course of a plant-hunting expedition in 1828.

Amaryllis belladonna arrived in England around the early 18th century, preceded by its arrival in Portugal, Spain and Italy as explorers from these countries traveled far and wide, returning home with new discoveries to please the wealthy patrons who paid for their expeditions. To Northern Europeans, the pride of the flower was connected to the pride of their southern cousins who introduced the flower to them.

When you are handed an amaryllis, reflect on this: to have pride in accomplishment is no shame, but be careful that the pride is justified. Stems that grow too tall often crack at the base.

Malus domestica

It may have been an apple that tempted Eve, but an apple blossom promises prettier things. 'I prefer you' said the one who sent their love a flowering sprig of apple. 'Fortune favors you.'

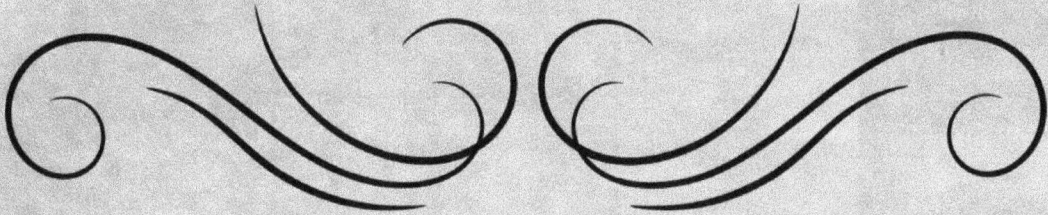

Asteraceae

In Latin, "aster" translates as "the star", a name referencing the tale of the goddess Virgo: by scattering stardust on the earth, she created starry flowers. The star-bright eye of the aster drives off evil influences in English and Mediterranean country magic.

Blooming to the very last of the killing frosts in the year, the aster is steadfast in its colors and its message. To receive an aster is to be told that you are faithful.

Temperance

Azelea

Rhododendron spp.

To carry an azalea bloom in the age of Queen Victoria announced that you supported the prohibition of alcohol. In fact, the soft flowers on the shrub's sturdy branches put folk in mind of emotional evenness of all sorts. One of the toughest of the newly introduced South American plants in Europe in the early 1800s, it needed less coddling than most emigres, and came to represent the good taste of wealth used sensibly. 'Be steady and sensible', so says the azalea.

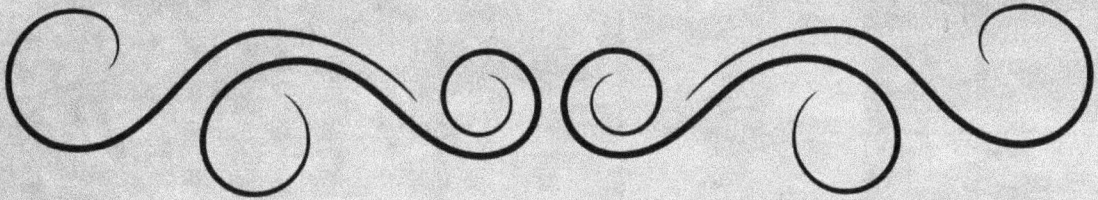

Innocence

Baby's Breath

Gypsophila

If roses are like bread to a florist, baby's breath is the butter. The fine, tiny blooms represent the innocence of small things in the mind of the floriographer.

Swiss-Russian botanist Johann Amman sent the first sample of the plant from St. Petersburg in 1736. The sample travelled all the way to Carl Linnaeus, the famed botanist who initiated the binomial classification system for plants. It was Linnaeus himself who dubbed the little plant gypsophila, meaning "chalk loving", referring to the conditions preferred by the baby's breath plant.

The sweet whimsy of baby's breath will always add a note of innocent pleasure to the world of flowers.

JUSTICE

BLACK EYED SUSAN

Rudbeckia

Named for the Swedish botanist Olaf Rudbeck, the black-eyed Susan takes its common name from an English song. People often mistakenly believe that the plant is named for its 'rude' appearance, and indeed it is direct and to the point.

Since the recognition of Rudbeck's work with a plant named for him was done in honor of all he had achieved, the writer Henry Phillips linked Rudbeckia with justice done and just desserts. When handing out black-eyed Susans, one sends a clear message: 'Let justice be done.'

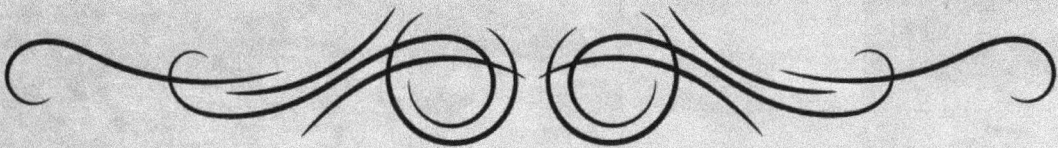

Gratitude

THANK YOU

BLUEBELL

Hyacinthoides non-scripta

In the 7th century, the Bishop of Aurelia fell gravely ill. The bishop was suffering from a terrible hemorrhagic fever, and blood stained his sheets. Fearing the possibility that he would infect others, he had sent his servants away and lay alone. Lying in bed, looking out the window in his home high on the hills outside of town, he could feel his life draining away.

As he lay making peace with his God, he saw a terrible vision: raiders approaching his town. High on the hill, he was the only one to see it.

In spite of his terrible frailty, the old man forced himself out of his sick bed and into his chapel, where he rang the bells. The sound alerted his flock to the threat, and the defenders raced to the city walls. After they won the battle, the flock found their dear bishop dead in the bell tower. But they found a new flower as well: a nodding blue bloom that sprang up wherever the dying man's blood had fallen. When the bishop was buried, thousands of these fascinating new blooms grew over his grave.

Since that day, handing another person a bouquet of bluebells expresses your gratitude. 'I am grateful for the good things you do in my life,' say these nodding blue columns on their slender stems.

Caution

BECOME

Begonia coccinea

The Begonia's symbolism of caution comes directly from the translation of the name. This unusual moniker for the plant was chosen by Charles Plumier, the Franciscan monk who discovered it. Plumier found fibrous begonias in 1690. On a search for medicinal plants in Brazil at the time, the good father named the plant after his favorite botanist and helpful patron, Michel Begon, the governor of Haiti. Unfortunately, Plumier passed away soon after the discovery and the documentation of the plant did not go further until the 1800s. When it reached Europe, its name was quickly recognized as the Latin word for danger. Soon it became the fashion to hand those in danger of embarrassment a begonia.

Indifference

CANDYTUFT

Iberis sempervirens

The name "candytuft" does not come from any relation to the sweet shop, but from the process of anglicizing the word "Candia". Known as Crete today, Candia was the homeland from which this plant was imported to England in Elizabethan times. The genus name, *Iberis,* comes from the Latin word for ancient Spain, where many native varieties grow.

Often called "Poor Man's Mustard", the seeds were used as mustard substitutes from the 1500s onwards. Candytuft was soon scorned as the mark of a palette that was indifferent to quality and would accept the inferior article.

'I could not care less,' is the message sent by a paramour delivering this flower. 'I am utterly indifferent.'

The Answer is No.

Striped Carnation

Dianthus caryophyllus

Legend has it that the striped or bicolored carnation came into being through a tragedy. In the Peloponnesian War, a maiden called Margherita gave her sweetheart Orlando a white carnation, which he carried with him as he entered the fray. Orlando was mortally wounded, and his blood spattered the white petals with red. To this day, this flower carries the message that a love or a relationship cannot continue.

In the interest of honesty, I must add that according to the more prosaic writing of Elizabeth Wirt, the answer to the reason for this plant's meaning may be much more utilitarian. In *Flora's Dictionary* she wrote, "If the White Carnation answers with a Yes, her inverse answers in the reverse."

The Answer is Yes

WHITE CARNATION

Dianthus caryophyllus

The carnation's name comes in part from the word "carno" or "flesh colored". In spite of the name, the pure white carnation's color is clean and bright. White carnations represent pure, platonic love.

Wearing a white carnation in the buttonhole symbolize luck and your willingness to be friendly. As a side note, red carnations in a buttonhole encourage a courageous nature.

"Yes, the answer is yes," the white carnation tells the recipient.

Endurance8

CHAMOMILE

Chamaemilum nobile

The name "chamomile" originally comes from the Greek word "Khanimimelon", meaning 'apple of the Earth', due to its low growth habit and sweet scent. In the Victorian mind, this old friend of the garden earned the appellation by its ability to survive even the most strenuous trampling and abuse.

Maud Grieve says this of chamomile in *A Modern Herbal* "When walked on, its strong, fragrant scent will often reveal its presence before it is seen. For this reason **it** was employed as one of the aromatic strewing herbs in the Middle Ages, and used often to be purposely planted in green walks in gardens. Indeed, walking over the plant seems especially beneficial to it.

"Like a camomile bed
The more it is trodden
The more it will spread."

Beauty

Cherr Blossom

Prunus serrulata

Victorian Europe took much of its symbolic understanding concerning the cherry blossom from what they called 'The Orient'. There was a fad in the Victorian Era for all things oriental, and in this time a great many images of handsome Oriental ladies with cherry blossoms in their hair and cherry trees around them were disseminated on the Continent, in England and all its colonies. Thus, in the European mind, the cherry blossom became associated with beauty. But what the Victorian ladies and gentlemen admired, they were still too ethnocentric to truly understand.

In Japan, "sakura", the cherry blossom, is an integral part of the culture. The flower is tied in Buddhist philosophy to the shortness of mortal life and all its fleeting beauties. The cherry blossom reminds us to enjoy the moment we are in, for it will not be here long.

Industry

RED

CLOVER

Trifolium pratense

One leaf for fame, one leaf for wealth,
One for a faithful lover,
And one leaf to bring glorious health,
Are all in a four-leaf clover
(Author Unknown)

This is the poem that explains why a four-leaf clover is lucky. But there's another saying too: 'The harder you work, the luckier you get.' Red Clover earned its appellation of the Flower of Industry from its utility when planted in compacted soil. Not only will the clover thrive on the worst soil, it will actually leave the ground better than before: aerating the soil, filling it with nitrogen and acting as a green manure in order to allow abused soil to regain its water- retentive properties.

Victorian farmers regularly planted fallow fields with red clover in order to fortify them. This is the plant to hand the industrious person in your life. 'Your hard work will bear fruit,' red clover promises.

Jealousy

Daylily

Hemoerocallis fulva

With blooms that survive only a day and wither in the evening, daylily already invites comparison with unfavorable human traits such as fickleness and a jealous manner in the mind of the floriographer. That is probably how it became connected with an ancient Greek story in the Western mind, despite its origin in China and late introduction to Europe in the 1800s.

The story goes that when Venus rose from the water and saw a lily on the day she was born, she became envious of the flower's charms. She couldn't stand its beauty and, feeling threatened, gave the flower a long pistil at its center in the hope that this would make the flower less attractive.

"Your inconstant attentions bring forth jealousy," states *Le Langage des Fleurs* in regard to this flower's message. Present it to another with care.

Fascination

FERN

Rumohra adiantiformis

Ferns and fascination go hand in hand. There's their growth habit, entrancing fiddleheads unfolding day by day. There's their native environment, shady nooks in forests where the fairies might still linger. There's their strange reproduction: for many centuries they were believed to have invisible seeds. Perhaps it's no wonder that some gentlefolk came down with pteridomania.

"Pteridomania"- an obsession with ferns- gripped amateur botanists beginning with the invention of the Wardian Case in 1829. With Ward's technology, George Loddiges built the world's largest green house and fern conservatory. But this glass creation needed patrons to earn its keep, so Loddiges put it about that the study and collection of ferns improved the brain and male virility. His friend Edward Newman published *A History of British Ferns,* which supported Loddiges' claims.

The book caught on, and so did the idea of ferns. Soon, live ferns were *de rigour* in houses rich and poor, and fern motifs were printed on anything that could be bought. Those who could afford to, spent vast amounts on importing the newest and rarest fern varieties for their Wardian cases, fascinated by their forms.

To hand someone a fern is to say, 'you fascinate'.

Friendship

FREESIA

Freesia alba

It appears that freesia was first brought to Britain from the Cape of Good Hope about the year 1850 as the species Freesia Refracta, which has sweetly perfumed white flowers. The plant is named after a physician and botanist from Germany by the name of Frederick Freese, a friend and student of Dr. Christian P. Ecklon, who named the flower after him.

It took some time to catch on, but today they are one of the most popular cut flowers. The lovely story of the freesia's naming became public as quickly as the plant itself, and as soon as it was introduced it became connected in the public mind with platonic love and trust between good friends.

Today there are over 1,400 species of cultivars in all the warm-spectrum shades, from white through deep mauve. Freesias of all colors symbolize trust and innocence, but there are several freesia colors that carry specific meanings:

White ~ purity and innocence
Multicolored ~ friendship and thoughtfulness
Pink ~ motherly love
Yellow ~ joy, renewal, and friendship

Deceit's

Foxglove

Digitalis Purpurea

It's said that in the Shropshire fields of England, this lovely flower earned its common name from the use the neighborhood pixies put it to. Farmers in the neighborhood believed that the moor sprites stitched the blooms into gloves for their friends the foxes, in order to disguise the raiding animal's tracks.

In those same Shropshire moors, the physician William Withering gave up on a patient suffering congestive heart failure and advised him to set his affairs in order. The professional was quite shocked when, six months later, he met his dying patient in the street, in perfect health. The man explained that he'd gone to a local goodwife for healing after he was given up for dead. While interviewing the woman and going over her remedies, Withering discovered one pretty purple flower with powerful cardiac effects. This plant soon yielded digitalin, one of the most potent cardiac glycosides known today. It immediately became a boon to medicinal men.

Soon after, it also became a boon to murderers. In the wrong hands, digitalin became an insidious poison, difficult to distinguish from natural heart failure in the hands of a careful criminal. Even in the skilled hand of the doctor, digitalin proved to be a terribly tricksome drug, and more than a few patients died when the dose proved fatally under or over dosed. The pretry purple blooms have caused fascination and death since time out of time, their cheery beauty masking their potency. Perhaps that is why one of its names is 'dead man's fingers' and another common name is 'dead man's bells'.

'I do not trust you,' says the one who presents you with a foxglove bloom, 'you are deceitful.'

Foolishness

Geranium

Pelargonium peltatum

'You are making a fool of yourself. It does not amuse,' the Ivy Geranium baldly tells its recipient.

We read in *Fifty Easy Old Fashioned Flowers* that "The first geraniums sent to Holland came from South Africa in 1609. By 1650, the plants were common in Europe, where they were grown for their beauty as well as for their fragrance."

So, why do we end up with one variety of geranium meaning such a negative thing, while the rest of the cultivars denote 'steadfast favor' and 'friendship' or even 'may I have your hand for the next dance?'

It·a n comes down to an argument between botanists. Linnaeus placed all geranium and pelargonium plant species in the genus Geranium. A few years later, the French botanist L'Heritier, noting that some geranium species of plants were so distinct that they should be in a different genus, formally transferred them from Geranium to Pelargonium. However, for reasons unknown, this change in names was not accepted by all botanists and garden writers.

With its low, trailing growth habit, Ivy Geranium most closely resembled wild geranium, which was now the only true 'geranium' in the botanic sense. People constantly mixed up the new nomenclature, and the wags in the botanic community soon nicknamed Ivy Geranium 'Fool's Geranium'. This garden spat was carried in many horticultural periodicals of the time, and it must have caught the eye of floriographers like Harry Phillips and Elizabeth Wirt, who immortalized the Ivy Geranium forever as the flower of the fool.

Helenium autumnale

Named for the tragic Helen of Troy, Helenium blooms just when most perennials are fading. Its sad namesake, both 'sinning and sinned against' as the old saying goes, has linked it with sorrow and with painful situations in spite of its lovely orange color.

History gives us a somewhat less poetic reason than legend for the association: helenium was often used to flavor snuff, a smokeless tobacco product made from ground or pulverized tobacco leaves. Originating in the Americas, by the 17th century, snuff was in common use in Europe. Snuff is inhaled or "snuffed" into the nasal cavity, delivering a dose of nicotine and a lasting flavored scent.

It was polite to snuff the product gently, but many a gentleman (ladies were discouraged) overdid the snuff and using it would be followed by an explosive sneeze and watery eyes. This earned helenium, so often used as part of snuff, the common name "sneezeweed".

Poetic or prosaic in reason, helenium states 'You have me in tears.'

HOLLYHOCK

Alcea Rosea

The accepted theory on this flower says that the crusaders brought back hollyhock seeds from the Middle East, and a little etymological research encourages this idea. "Hoc" is Anglo-Saxon for "mallow", and "holly" is a common corruption of "holy". With that in mind, the name can be read 'the mallow of the Holy Land'.

In Western European magic, hollyhock seed pods encourage fertility, and in Eastern European magic it is a protective charm to ensure a happy home.

At first blush, it seems odd to put down the quintessential cottage flower as the flower of ambition, until you see the way the plant seeds. Prolific seeds are dropped in the fall, and soon two or three hollyhocks become a great drift of them, standing tall and proud. Their height, their drive and their showy blooms have earned them the honor of standing for healthy ambition.

This is particularly appropriate as a gift for an ambitious woman. Hollyhock flowers and seeds were a common gift for women embarking on important ventures: moving into their new home after marriage, having their first child, moving to a new country.

Today, it would be equally appropriate to gift a lady starting a business or being promoted with a bouquet of hollyhocks, to let her know how thrilled you are that she's reaching for the stars.

Forgiveness

HYACINTH

Hyacinth

The name "Hyacinth" is derived from the name "Hyakinthos", and the story of a tragedy. In Greek mythology, Apollo, the Greek god of the sun, and Zephyr, the lord of the west wind, were both in love with the same young man who went by the name Hyakinthos. Both vied for his attentions, bringing him gifts and sweet poetry. Hyakinthos, however, was a bashful young man, and did not return the affections of either god with certainty. At one point, Apollo decided to teach Hyakinthos how to throw the discus, taking the opportunity to hold the gorgeous young man close. Zephyr grew so angry that he blew a gust of wind in Apollo's direction, which sent the discus hurling back in the direction of Hyakinthos. It struck him on the brow, killing him instantly. Apollo, brokenhearted, made the body into a beautiful blossom, naming the flower Hyacinth in honor of the boy.

Originally from the Mediterranean, Iran and Turkmenistan, the plant arrived through the offices of the German doctor Leonhardt Rauwolf, who collected samples of hyacinths when he visited Turkey in 1573. By the early 18th century, 2,000 cultivars were available and beloved throughout Europe.

There are three appropriate times to give hyacinths: when you are wishing good luck on a fresh start in someone's life, when you are celebrating Spring, and when you are saying 'I'm sorry I hurt you. Please forgive me.'

Eloquence

Iris

Iris germanica
(and a few others)

Named for the Greek goddess of rainbows and communication, Iris comes in over two hundred varieties and a vast array of colors.

It's a flower with a long history: King Thutmose brought the flower to Egypt from Syria among his spoils of war before Rome was in its prime. It became a symbol of intelligence when king Clovis of France, desperately trying to think of a way to get his army out of the trap the Goths had set in A.D. 496, saw a yellow iris growing in a river and educed from it that the water was low for the time of year, and shallow enough to cross. His clever realization saved the lives of his army.

Some say this was the origin of the *fleur de lis,* Clovis having claimed and stylized the iris as his emblem in gratitude. The story's truth is debatable (the lily also has a solid claim to being the true *fleur de !is)* but it's certainly the type of story that people remember. Thus, the flower became a symbol of bright creativity and intelligence, with particular colors denoting specific aspects of a bright soul:

Purple ~ Your eloquence is a joy
Blue ~ I love your cheerful determination
Yellow ~ I admire your clever ideas
White ~ Your clear thinking is admirable

45

Lavandula officinalis

If a suitor had reason to believe he wasn't the only one vying for a lady's affections, a gift of lavender would let her know of his devotion, but also his mistrust. The English word "lavender" is derived from the Old French "lavandre", which in turn is derived from the Latin *lavare*, meaning "to wash". The botanic name Lavandula and other European vernacular names for the plant are considered to be derived from this root. The plant lived up to its name and was commonly used in the washroom to scent clothing and prevent attacks by moths and other insects on the cloth. But in more refined settings, lavender implied anxiety over one's own worth. The 17th century song "Lavender's Blue" says it well in its wistful verses:

> Lavender's green,
> Dilly dilly,
> Lavender's blue.
> If you love me,
> Dilly dilly,
> I will love you.

Medicinally used to soothe anxiety and insomnia, this flower tells its recipient, 'The giver is not sure of you or the proper way forward. Please provide reassurance'.

atred ORANGE LILY

Lilium bulbiferum

To hand another an orange lily is to say, 'I hate you', though sources differ on the reason.

In AD 800, the pope supposedly presented the emperor Charlemagne with a banner showing golden lilies on a blue background. Thereafter, the *fleur de lis* was used in France to signify divine right. Over time and through wars, this symbol came to represent all that other countries disliked about the French: their haughtiness, their disdainful manner and their cruelty. The worst of French criminals were branded with a *fleur de lis* in prison to warn the public of their nature, and so the symbol became associated with French villainy in the European mind.

Henry Phillips suggests in *The Floral Emblem* that the orange lily is a "debased and licentious" version of the pure white lily. It follows in symbolic logic that if the white lily stands for purity, the overdressed orange lily stands for tawdry vices worthy of hate.

Or perhaps Victorian ladies and gents really *did* long for a way to say 'bugger off!' in their genteel society. Tossing the annoying party an orange lily certainly did the trick!

Forgetting

Laudanum

Lotus

Nelumbo nucifera

An ancient flower with deep religious connotations in Asia, India and Africa, the lotus came to Europe as a symbol of spiritual perfection, due to its growth habit of rising pure and white from the mud in which it grows.

But in Europe, it got tangled up with the Greek story of the lotus eaters in the Odyssey ("lotus" being a mis-translation of the text. The folk were actually eating poppies, which makes much better sense in context). With this connotation, this symbol of noble spirit in the East · became a symbol of forgetfulness in the West.

Of course, there are times when forgetting is a kindness rather than a vice. The lotus is an ambiguous flower which can mean 'please forget what has passed', 'I hope time eases your painful memories', and 'all things pass in time'. It can also stand for 'I have forgotten the past'. Look to the flowers delivered alongside the lotus for context.

Nigella damascena

The common name for this plant stems from its appearance, with many wispy bracts creating a mist of green encircling the mauve blooms.

A native of southern Europe and northern Africa, it's a common cottage garden flower and even a vegetable garden flower. It was.commonly used as a substitute for expensive nutmeg, its similar flavor making a pleasing addition to the table.

'You perplex me', the Nigella flower states to its recipient, 'I am bewildered'. In context, it also carries the connotation of 'please, make your intentions clear?'

Nobility

MAGNOLIA

Magnolia grandiflora

First cultivated in China, the magnolia's modern name was given to honor Pierre Magnol, the director of the Montpelier Gardens during the time of the tree's introduction into European society.

Since the day of its arrival in Europe, the magnolia has been seen as a symbol of nobility. At times referred to in period literature as a symbol of graciousness, good breeding or quality, nobility was often a mere description of socio-economic status which the Victorians mistakenly tangled up with the state of the person's morals.

But there is a deeper quality that makes a truly noble character: self respect, care for others and a wish to do right by them, determination and poise shape a person's character far more than whether they can afford a teapot of silver or china.

'I think you are truly noble', the magnolia tells the recipient. How they conceptualize nobility is their affair.

Tropaeolum majus

With the lovely name of nasturtium, literally translated from Latin as "the twist of the nose", how can this bright flower help but be the trickster of the green world? These peppery, fire colored flowers were capable of various meanings. 'I was only teasing', 'you jest' 'you make me laugh' and 'don't take this seriously' were the main thoughts imparted.

However, the jester should take note: a joke may seem far funnier to the one making it than the one it is made upon. Have a care lest your laughter turn cruel.

Arriving on the shores of Europe from its Peruvian mountain home in 1574, nasturtium wasted little time in colonizing kitchen gardens and informal beds, where it became a favorite spicy salad green, delighting both the eye and the tongue.

This is the bloom that will remind its recipient to smile.

Nerium oleander

The Latin name for this plant, "nerium", comes from the word for "wet" or "fresh", and the leaves are forever glossy. Stroke them at your peril though, for the plant's leaves are full of potent cardiac-affecting glycosides. It's been known to poison men, cattle and horses who associated with it, and throughout the Mediterranean it was once planted as a hedge to warn both man and beast away.

Even so, elegant ladies wore oleander in their hair to attend the best balls, smitten with its beauty. At least one case in 1880 reports a woman who, upon finding that she was ruined by a scandal that came to light at just such a ball, used the oleander blossoms adorning her hair to brew a strong tea, and committed suicide in her despair.

Be wary when the oleander is handed to you. Trouble is not far off.

Thought

PANSY

Viola tricolor var. Hortensis

The word "pansy" comes to us from the French "pensee", the word for "thought". To pick a pansy while thinking of your love, it was believed, was to make them think of you in turn.

In modern lexicon "pansy" is often used as a pejorative connoting weakness. But, if you are a gardener, you will know that the pansy is one of the first flowers to raise its face to the sun in the spring and one of those blooms still bright with color when the frost has killed most other plants in the fall. It is easy to step on a pansy, but the species can endure. It is easy to kill an individual, but ideas have a life of their own.

Give someone pansies to let them know that they are in your thoughts.

Compassion

PEONY

Paeonia lactiflora

"The roses here are as big as cabbages," wrote Marco Polo during his time in China. In fact, what he was viewing in wonder was not a rose, but a peony. In China, the peony is the national emblem, and the flower is honored as the flower of everlasting life. Peonies are able to live for a hundred years, blooming every spring in startling color and voluptuous shape.

Since they took great care to propagate, they were very expensive and thus were flowers of prosperity in the East. As the imperial.symbol, the flowers were spread across the country as different emperors moved their courts. They reached Japan around the beginning of the 8th century and were brought to England by the Romans in the year 1200.

The softness of the flower and its generous form led the Victorian writers to link it with soft hearts and platonic compassion. It's also linked to the story of a young man called Paeon, who Zeus took pity on and transformed into a flower in order to save his life when he was being hunted by a murderous teacher.

'I care for you and your condition,' murmurs the Peony.

Change

Pimpernel

Anagallis arvensis

Red here and blue there, ever changing everywhere. This is the Scarlet
Pimpernel, whose nature is change. Not only will the plant change color
depending on the part of the country and the amount of light it receives,
the plant is also photosensitive and will close its flower petals up when
the day is grey, making it even more a symbol of changing times.
It's so linked with change that it was given as the name of one of the
most famous spies and disguise artists in literary history, the Scarlet
Pimpernel.

> They seek him here, they seek him there,
> those Frenchies seek him everywhere.
> Is he in heaven or is he in hell?
> That damned elusive Pimpernel
> *(Author Baroness Orczy, 1905)*

But how is it possible for one plant to produce flowers of two colors,
one in sun and one in shade? The *Daily Telegraph* gives us the answer.
Ken Thompson writes, "Spanish researchers mapped red and blue
plants across Europe and found that by far the best predictor of flower
colour was hours of sunshine. The sunnier it is, the more likely are
blue flowers, so you can see right away why they're red here and blue
in southern Spain. When they grew both forms under experimental
conditions, the blue form also tended to do best when it was dry or
sunny."

To hand someone a scarlet pimpernel is to promise a coming change.

Cydonia oblonga

Some say that the fruit of the forbidden tree in Eden was a quince rather than an apple. With its burning red blossoms and golden fruit, the quince put the Victorian intellect in mind of red velvet, gold, and all the enchanting and dangerous things associated with them.

In Greek myth, quince is the fruit Eris gave to Paris when she requested that he offer it to the most beautiful of three ancient Greek Goddesses: Athena, Artemis and Aphrodite. The promise Aphrodite gave him in order to secure the precious fruit for herself was the cause of the treacherous ten-year Trojan War. If Paris was to select Aphrodite as the rightful bearer of the prize, the goddess would make sure he would have the most beautiful mortal woman of the time, Eleni. This name we have anglicized as "Helen". So, 'the apple of discord' was no apple but in fact a rather troublesome quince!

The tree is native to rocky slopes and woodlands in Asia, Turkey, northern Iran and the Middle East, though today it thrives in a variety of climates and can be grown successfully as far north as Scotland. The quince's botanical name, Cydonia oblonga, derives from Kydonia on the island of Crete.

'I am tempted,' the quince blossom says when it is offered, or at times it can warn 'this situation is a dangerous temptation'.

Rosa spp.

(the species names would fill a page.)

You could fill a book with all the things there are to say about roses, and more than one writer has. To begin with, you could answer the question 'what does a rose mean?' with another question: 'which rose were you considering?'

The genus Rosa has some 150 species spread throughout the Northern Hemisphere, from Denmark to Northern Africa and Alaska to Mexico. The number of cultivars is somewhere in the thousands, equal to the number of tales involving this blossom.

The Latin expression *sub rosa* referred to confidential information, and in ancient Rome a wild rose was placed on the threshold of a room where confidential matters were being discussed. A wild rose can still denote secrecy today.

The red rose has another story. In Arabic legend, all roses were white until the nightingale met a beautiful white rose and fell in love. At this stage, nightingales were not known for their melodious song. In his devotion, the nightingale was inspired to sing for the first time. Yearning for his darling, he pressed himself to the flower and the thorns pierced his heart, coloring the rose red forever.

During the 15th century the rose was used as a symbol for the factions fighting to control England. The white rose symbolized York, the red rose symbolized Lancaster, and as a result the conflict became known as the "War of the Roses". Only Henry's Tudor rose, red and white at once, brought peace to England, which makes a Tudor rose a fitting gift after a fight.

Have a care about your culture when you're choosing roses as well: to give a yellow rose in England is to say 'you are inconstant'. This was brought about by a breeding issue in the English tea roses solved by M. Pernet-Doucher only in 1900. Before his time, yellow English roses tended to revert to pink within a generation.

In America, the story was entirely different. The tough native roses of America, long used by the tribes of the Pacific Northwest for winter sources of vitamin C, hybridized readily with *Rosa foetida,* a yellow rose native to the Caucasus mountains, to produce Harison's Yellow. It first appeared in the garden of George F. Harrison, an attorney in New York City in 1824. William Prince, nurseryman of Long Island, took cuttings and began to sell it in 1830.

A shrub rose with canes five feet tall, Harison's Yellow is happy growing almost anywhere, and it soon did. A great many European-descended women purchasing bare-root canes before they left to 'settle' the American west were sold Harrison's Yellow. Kept moist during the journey, the canes would have developed roots and been ready to plant when the women reached their new homes. They say that you can trace the Oregon Trail by following the bright yellow flowers of Harison's Yellow.

A rose by the door became a symbol of a home with a good woman in it, a .home that pledged a future of more than hardscrabble toil. To see a yellow rose was to understand there would be kindness and a welcome for a weary traveler, and so in America the yellow rose became the mark of friendship and good cheer.

The sentiment has been immortalized in American song through the words of "The Yellow Rose of Texas". This song was, interestingly enough, about Emily D. West, a free black lady who loved Colonel James Morgan in 1835. In the attack of Santa Ana, Morgan lost West and was never able to find her again. Legend tells us he wrote the song in *memory* of her.

As a general guide to this complex blossom, follow these rules of thumb when giving gifts.

Rose (orange) ~ Fascination

Rose (peach) ~ Modesty, gratitude, appreciation, admiration, sympathy

Rose (pink) ~ Grace and admiration,
young or filial love

Rose (purple) ~ Enchantment

Rose (red and white) ~ Unity, end to hostility

Rose (red) ~ Romantic love

Rose (white) ~ Purity, platonic love

Rose (yellow) ~ Joy, gladness, freedom, jealousy,
infidelity (mind your culture!)

Rose (burgundy) ~ Sexual passion

Rosebud ~ Youth and beauty

Dog Rose ~ Simplicity, secrecy

MEMB ROSEMARY

MEMENTO MORI

Rosmarinus officinale

"There's rosemary, that's for remembrance; pray, love, remember." Perhaps one of the most famous scenes concerning flowers there ever was, Shakespeare's scene of floral language in Hamletincludes, among other things, the message of Rosemary: remember.

Another Mediterranean native, this evergreen shrub is most often seen growing by the ocean in the wild: its Latin name translates as "dew of the sea".

Rosemary was used- rather ineffectively-throughout Europe during the Black Plague to ward off sickness. At the time people believed in its ability to protect against the plague, and consequently the price soared.

Although simply carrying rosemary would do little to prevent sickness, modern research has confirmed its antibacterial, antiviral, and antioxidant abilities. Several notable Roman figures wrote about rosemary's ability to support memory as well. Pliny the Elder and Galen both wrote about the benefits of rosemary, and Dioscorides, the author of *De Materia Medica,* also expounded upon its healthful benefits. Several authors in the 16th century supported rosemary's ability to treat a wide range of mental ailments.

The *Grete Herbal* reads, "Rosemarie - For weyknesse of ye brayne. Against weyknesse of the brayne and coldenesse thereof, sethe rosemaria in wyne and lete the pacyent receye the smoke at his nose and keep his heed warme."

Throughout its history, rosemary has been a flower of memory and remembrance, appearing both at weddings to encourage marital fidelity and at funerals to symbolize undying memory of the dear departed.

Involved in remembering by referring to the plant:

A seventeenth century song records the pain involved in remembering by referring to the plant: "Are you going to Scarborough Fair:/ Parsley, sage, rosemary, and thyme./ Remember me to one who lives there./ She once was a true love of mine."

"I will remember." a sprig of rosemary tells the recipient. Whether that is good or ill depends entirely on the people involved.

Courage

SNAPDRAGON

Antirrhinum majus

Derived from the Greek words "ant," meaning "like," and "rhin," meaning nose, the snapdragon's botanical name, Antirrhinum, is a fitting description of this snout-shaped flower.

The common name for this colorful bloom comes from the snap it makes when the sides of the 'dragon's mouth" are gently squeezed. In this collection, snapdragon is one of the true European natives, growing wild in Southern France among its toadflax cousins and standing tall in the ruins of Roman villas.

The common name of this flower and its proud spires of color linked it in the European mind to the concept of knight's lances and courageous knights fighting dragons. This is especially true in England, where Saint George is an integral part of the national story.

'Have courage', the gift of snapdragons encourages, 'the battle can be won'.

SNOWDROP

HOPE

Galanthus nivalis

Carey wrote, "This plant appears in the middle of the snows clothed in its charming dress like an imprudent and coquettish nymph, who, in the winter, trembling with cold, dresses herself in spring attire."

Far from imprudent, snowdrop has long been seen as the symbol of resolute hope in adversity. Also called hopewort and the "Fair Maid of February", they are the harbingers of spring in the depths of winter. Considering how hard the frozen ground is when they bloom, and how delicate the snowdrop's stems, their presence seems miraculous even today.

"The snowdrop, in purest white arraie, First rears her hedde on Candlemas daie", reads the English tome *Floral Calendare* dating from around 1500. February the Second, Candlemas Day, was also known as the Feast of the Purification, commemorating both the presentation of the infant Jesus at he Temple and the Virgin Mary's purification. As the name "Candlemas" suggests, candles are traditionally lit at the service that marks these events, and it is likely that the practice of kindling flames and lighting candles on this day originated in pagan times. In the pre-Christian Celtic calendar, this day was marked as Imbolc, "milk day", and was the day sheep began to drop their lambs and provide much needed fatty milk to be made into butter and cheese just when winter food supplies were beginning to run low. In spiritual terms, it was seen as the day the goddess Brighid began to travel the land in a green mantle and drive out the Calliech Bhuer, queen of winter. Snowdrops were believed to spring up where the heat of Brighid's feet melted the snow, a story still told by English, Scottish and Irish country folk well into the 20th century (after Brighid was suitably covered in the mantle of a saint, of course.)

'Better things are coming', the snowdrop whispers, 'have hope'.

Strife

Teasel

Dipsacus sylvestris

Years ago strangers remarked about the lazy farmers around Skaneateles allowing their field to be overgrown with thistles.
—Henry W. McLaughlin, "This History of the McLaughlin Family"

Of course, those strangers did not understand what they saw in the fields of New York. Those thistles - called teasels - were in demand throughout the world at wool mills. It turned out that soil and climate conditions in the Skaneateles area were ideal for teasels. For many years this "weed" was the town's most famous- and unusual - cash crop.

The man owning most of this crop, often called "The Teasel King", was James McLaughlin Jr. Truth be told, other gentlemen also laid claim to the title. An article by Charlotte Coffman in the April 2000 *Textiles and Apparel* newsletter said the first Skaneateles teasel grower was William Nipper. Another article said Skaneateles businessman Walter Hamilton Kellogg (1860 - 1934), called himself "America's Teasel King". Suffice it to say, teasel was making money and enemies.

Dr. John Snook bought and planted the first teasel seeds in 1833. He had settled in Skaneateles because he felt it had the ideal soil for teasels, which were used in the production of woolen cloth. The heads of cultivated teasel are used for wool "fleecing", raising the nap on woolen cloth, and no machine has yet been invented which can compete with teasel in its combined rigidity and elasticity. Dr. Snook found an abundance of limestone in Central New York, and limestone soil produces the strongest teasels. Thus was an industry born.

Teasel may have earned its association through what it does, ripping and pulling at woolen fibre until it is forced into smooth obedience, or it may have been its very appearance that caused the association. In Carey's *Language Of Flowers* we read "the flowers of the Fuller's teasel are set with long sharp thorns, the whole plant having a repelling appearance. It is, however, useful and beautiful, the drapers employ it to paint their stuffs, which has given it the vulgar name of Fuller's teasel."

Today plenty of people hate teasel in America and Europe for its invasive abilities and terrible habit of choking out every other species in a field. Few people like a field of thorns.

Few people like a field of thorns.

The teasel can send a handful of messages, none of them pleasant. 'There is strife in this situation', 'You irritate me,' 'you are a misanthrope' or 'I really do not like you,' are all possible meanings.

Oh, and watch out for the physical thorns, too. You've been warned.

passion

TULIP

Tulipa gesneriana

To the modern eye, the tulip is the quintessential pretty spring flower. But in its travels to cozy cottage gardens, this was a flower that inspired passion, avarice, the winning and losing of fortunes ... even madness.

The history of the tulip can be told in the history of its name. In Turkish, the flower's name was "dulband", anglicized as "turban": the flower could have been translated as something like "gentleman's headdress". Instead of translating it, the man who brought it to Vienna, one ambassador to Constantinople by the name Ghiskain de Busbecq, tried to pronounce it ... and failed. He presented it to King Ferdinand the First as "the tulipan". The sultan Sulyman the Magnificent, he explained to his liege, had presented them as a token of his esteem.

In 1559, the Swiss botanist Konrad Von Gersner published a description of the plant in Germany, earning a taste of immortality through the inclusion of his surname as part of the plant's moniker.

Two years later a poor botanist tried his hand at selling tulip bulbs in the Netherlands. Unfortunately, he was robbed, and his Dutch tormentors sold the bulbs instead.

This is when things began to get a little wild.

Between 1634 and 1637, the enthusiasm for the new flowers triggered a speculative frenzy in the Low Countries which we now call the tulip mania. This was partly due to their amazing variety. On account of a mutagenic virus, the flowers could turn up with fascinating and unexpected eccentricities in every generation: winged petals, frilly fringes, striking stripes and even flames; red marks on white backgrounds that were prized.

By 1634, the entire country was mad for tulips. Tulip bulbs became so expensive that they were treated as a form of currency, or rather, as investment ventures. Everyone wanted to buy the newest, the rarest and the most wonderful cultivar and sell it at enormous profits. This environment spawned many strange and terrible things. Thefts were done. Throats were cut.

One man is recorded as mistaking a tulip bulb for an onion and eating it with his lunch; he was charged of felony destruction and thrown into jail, where legend says he went mad thinking of the fortune he'd obliterated.

Over the course of three years the frenzy eased, but by that time the flower's prominence throughout Europe was well and truly entrenched and the tulip mania was never truly forgotten.

'I desire you with a passion', the tulip tells you when it is offered, 'I yearn for you'.

Have a care when you accept the tulip.

Bouquet Recipes
Ideas for any occasion.

*Note on substitutions: If tree flowers are out of season, the leaves of the appropriate tree can be used as a frame for the bouquet to send the same message.

The Beginning Bouquet

Give when welcoming a new neighbor into town,
a new child or a new venture into the world.

Ingredients:
Pink freesia (parental affection), white roses,(purity and luck) white carnations (luck) blue Iris (cheerful dedication) or aster (dedication and fidelity), baby's breath (innocence) and hyacinths (new beginnings)

Substitutions and Situational Suggestions:
For a housewarming, choose yellow and white or multi-color freesia rather than pink, and add apple blossoms if you can get them. Add white Iris. For the opening of a project, a business or a venture, use the multi-color freesia, add red clover and take out the baby's breath. Ivy is commonly added to intimate trust and fidelity to the situation, and thyme and marjoram both encourage light hearts and good spirits. When freesia's not available similarly colored roses are acceptable replacements. Just don't add red roses.

The Boudica Bouquet

Give to the wonder women in your life who are achieving great things

Ingredients:
Hollyhock (healthy ambition), multicolored snapdragon (courage), yellow freesia (joy), yellow and/or white iris (admiration of ideas), peach roses (admiration, respect), chamomile (endurance), aster (fidelity)

Substitutions and Situational Suggestions:
If you are good friends, add white and/or yellow roses to the mix. If the gift is related to business, add red clover. If this person does social justice or legal work, add black-eyed Susans to honor their work for justice. If this is a family member, add pink roses or pink freesia. Oak leaves could be used to frame the bouquet and connote strength, and thyme to intimate good spirits.

The Brilliant Bouquet

Give to honor academic achievement

Ingredients:
Yellow, white and blue irises (intellect, logic and eloquence), white amaryllis (pure or justified pride) white carnation (luck), fern (fascination) framed in magnolia leaves (nobility)

Substitutions and Situational Suggestions:
If you are good friends, add white and/or peach roses to the mix. If this is a family member, add pink roses or pink freesia. If this is a graduation, add snapdragon or gerbera daisies for victory and white amaryllis. Add chamomile to underline how hard they worked.

The Bugger Off Bouquet

For the person you truly despise and never want to deal with again.

Ingredients:
Ivy geranium, (foolishness), teasel (misanthropy) foxglove (deceit), white candytuft (indifference), and orange lilies (hatred)

Substitutions and Situational Suggestions:
White candytuft can be replaced with yellow and/or striped carnations (You have disappointed me, the answer is no). If the person is being rejected on the grounds that they're overbearing or harassment is involved, replace foxglove with red amaryllis or sunflower (haughtiness). If the recipient is a rejected suitor or ex-partner, make sure to include striped carnations. Foxglove can be replaced with yellow roses if you really must, but Americans be aware that it usually reads as friendship in the U.S.

The Benediction Bouquet

Give in times when congratulation or a lauding of talents are appropriate.

Ingredients:
Bluebell (gratitude), white freesia (friendship) peony (compassion) framed with magnolia leaves (nobility) and rosemary (remembrance)

Substitutions and Situational Suggestions:
Daisy fleabane can be added or substituted as another way to say 'thank you'. If magnolia leaves are not available, purple-leaf sage (grateful friendship) is a good substitution.

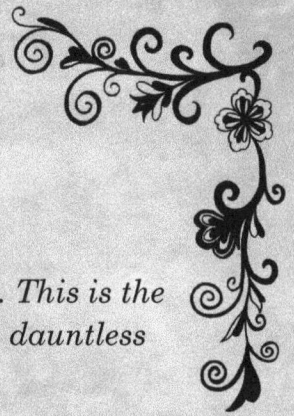

The Bedevere Bouquet

~

Give in honor of one doing sterling work under demanding conditions. This is the bouquet for diligent teachers, hard working secretaries or colleagues, dauntless friends and all others fighting the good fight.

Ingredients:
Red and white amaryllis (deserved pride), white carnation, white freesia (friendship), chamomile (endurance) peach roses (admiration), red snapdragon (victory) and magnolia (nobility)

Substitutions and Situational Suggestions:
Symbols of your particular relationships (pink roses for family, pink freesia for parents and children, red roses for romantic loves) can be added. Add situationally appropriate symbols such as black eyed Susan for justice or blue iris for a speaking achievement. If chamomile is not available, frame the bouquet in a spray of pine to give the same message.

The Bewildered Bouquet

~

Use when asking the recipient to make their position clear.

Ingredients:
A center of peony (compassion) should be framed with love-in-a-mist(confusion), lavender (doubt) and snowdrops (hope).

Substitutions and Situational Suggestions:
This is an extremely versatile bouquet. A wistful lover can add a single pink rose and a few begonias to denote 'I would like to show my affection, but I'm tentative.' Adding daylilies intimates 'take pity and tell me if there's someone else', a lotus would make it 'are you forgetting me?' begonia would soften the message to 'let me know, I'm unsure,' while adding oleander would sharpen the message into 'is this situation dangerous?' Adding helenium denotes 'you're bringing me to tears.'

The Big Kid Panties Bouquet

A bouquet to tell someone "I love you but you need to get your life together."

Ingredients:
Surround a single white peony (compassion) with pimpernel (change) azalea (temperance) hyssop (cleansing) pansy (thought) and hyacinth flowers (forgiveness) framed in lamb's ear leaves (support)

Substitutions and Situational Suggestions:
A lot can be added to this bouquet. For family or romantic partners add pink/red roses to underline the continuing love, white for a friend. If the person's indulgence is the reason for the message, add nasturtium and grape leaves or hops to say 'okay, you had your fun, enough.' Add willow and lemon balm for an emotionally-related situation to denote 'let's be level headed and work on this.'

The Bottom Bouquet

Give to someone who's making an ass of themselves as an encouragement to desist.

Ingredients:
Nasturtium (fun), ivy geranium (foolishness) helenium (tears) azalea (temperance) quince (temptation) and striped carnation (the answer is no) send a very clear message: knock it off!

Substitutions and Situational Suggestions:
Add grape leaves or hops if this is an indulgence-related issue. Add hyacinth to say 'you're an ass but I'll forgive you'. Adding teasle will make a much more blunt message along the lines of 'knock it off you ($)#!'

The Be A Dear Bouquet

Give to gently let an admirer know that they're cared for as a friend,
but not as a romantic interest.

Ingredients:
Yellow freesia or forsythia (good will,) azalea (temperance) striped carnation (the answer is no) and white roses (platonic love) should be framed in baby's breath (innocence)

Substitutions and Situational Suggestions:
I wouldn't change a great deal about this bouquet, as it is a delicate balance of negation and good will. If you feel badly about the situation in question, add hyacinth to show your feelings and ask forgiveness or cherry leaves and blossoms to say 'it was fun while it lasted.'

The Breakup Bouquet

Give to comfort a friend who's just gone through a breakup or divorce.

Ingredients:
White roses and peonies (compassion and platonic love) should be framed in pink yarrow (heartache) or helenium (tears) and willow leaves (serenity) or cherry(beauty is fleeting.) Add lemon balm for clarity and hyssop for cleansing.

Substitutions and Situational Suggestions:
If you want to reference the situation, daylily, oleander and orange lily are all good choices for the wronged party to let them know you sympathize with the crime against them. If you'd rather focus on the person than the problem, include lotus(forgetting) yellow roses or freesia(friendship) and aster(fidelity) to let them know that this will pass and they have friends.

The Bull-Headed Bouquet

Give to say, "let's agree to disagree. I love you regardless."

Ingredients:
Tudor roses (peace between old rivals) with lotus (let's forget) and baby's breath (innocence)

Substitutions and Situational Suggestions:
Add flowers denoting your relationship: pink roses for family, red roses for lovers, white or yellow roses for friends. If you feel it's needed, hyacinth can stand for forgiveness asked for or received, and bluebell will do the same for gratitude.

The Business Bouquet

Give to show a business associate or a friend in the business world your regard.
Especially appropriate when closing or negotiating business deals.

Ingredients:
A spray of hollyhocks(ambition) and white iris(intelligence)should be surrounded with pink and white carnations(luck and joy, the answer is yes) and red clover(industry) embedded in oak leaves(strength)

Substitutions and Situational Suggestions:
White roses would be appropriate for close associates. If the process was arduous, underline the good work with chamomile. If related to the law, add black eyed Susans.

The Blushing Bouquet

*Give to someone to signal your romantic interest when
you're too tongue-tied to say it aloud.*

Ingredients:
English daisy(innocence), pink roses(gentle love) and snowdrops(hope) should be framed with fern(fascination) to send a very pretty invitation without the need for words.

Substitutions and Situational Suggestions:
Red roses will make the gesture a little more direct. If you're feeling really bold, add tulips for passion or quince for temptation, making the meaning decidedly risque.

The Beloved Bouquet

Give to show your romantic love.

Ingredients:
Red roses (passionate love) should be combined with orange flowers (prosperity) apple(fortune) and magnolia(nobility) to denote your abiding love.

Substitutions and Situational Suggestions:
Add tulips for passion or quince for temptation if you feel like spicing this bouquet up. Add aster to remind your beloved of your fidelity.

The Beloved Friend Bouquet

Give to show your love for your friend.

Ingredients:
Peach, yellow and/or white roses (friendship and admiration) can be balanced with yellow freesia(friendship) and filled out with asters(fidelity) and nasturtium(fun)

Substitutions and Situational Suggestions:
This is another bouquet I wouldn't adjust much. White Irises could be used in place of roses or freesia for their friendly connotations. Peonies could be added for mutual compassion.

The Best Wishes Bouquet

Give to those in hospital or going through difficulties.

Ingredients:
A peony(compassion) should be surrounded with snapdragons (courage) chamomile(endurance), white yarrow and ginger flowers(health) pansies (we're thinking of you) and snowdrops (hope) embedded in oak(strength) or laurel(success) leaves.

Substitutions and Situational Suggestions:
In the case of illness, add thyme, feverfew and hyssop for health and cleansing. Flowers denoting your relationship always work.

The Beloved Departed Bouquet

Give to those suffering a loss.

Ingredients:
A white lily(pure love) should be combined with lemon balm(clarity), thyme(strength), rosemary(remembrance) white roses(friendship) and yew sprigs(death and rebirth) surrounded with cherry leaves (mortal beauty, swiftly passing)

Substitutions and Situational Suggestions:
Loss is extremely personal, so be careful of the connotations you give and the other person receives. If you feel it appropriate, lotus can be added to mark the fact that the pain will pass. Peonies could be used to show your compassion for those in mourning, but that's sometimes seen as tacky. Alternatively, white asters could be used to denote eternal fidelity. Yellow roses and white freesia are appropriate for friends. Flowers denoting the person's qualities and life achievements would be an elegant gesture.

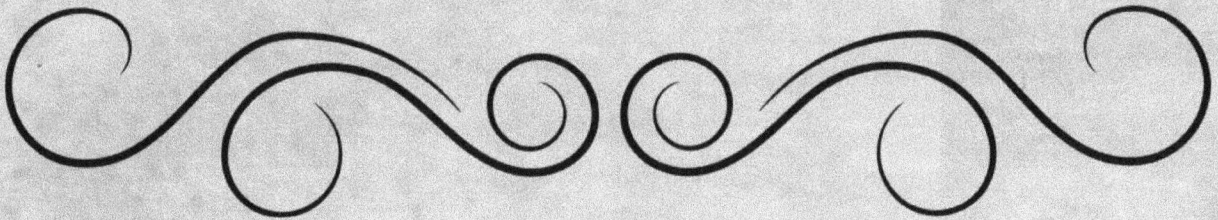

Further Associations

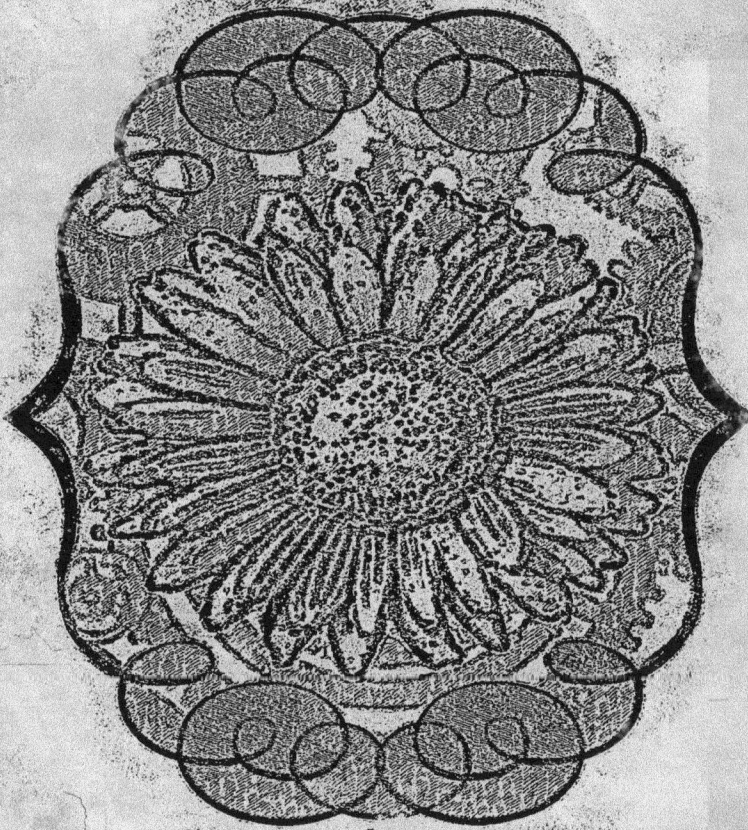

Birth Month Flowers

January ~ Carnation

February ~ Violet, Iris

March ~ Daffodil

April ~ Sweet Pea, daisy, peony

May ~ Lily of the Valley

June ~ Rose

July ~ Larkspur, Delphinium

August ~ Gladiolus, dahlia

September ~ Aster

October ~ Calendula

November ~ Chrysanthemum

December ~ Narcissus, poinsettia, holly, or paperwhite

Wedding Anniversary Flowers By Year

1st Anniversary ~ Carnation

2nd Anniversary ~ Lily of the Valley

3rd Anniversary ~ Sunflowers

4th Anniversary ~ Hydrangea

5th Anniversary ~ Daisy

6th Anniversary ~ Calla Lily

7th Anniversary ~ Freesia

8th Anniversary ~ Lilac

9th Anniversary ~ Bird of Paradise

10th Anniversary ~ Daffodil

11th Anniversary ~ Tulip

12th Anniversary ~ Peony

13th Anniversary ~ Chrysanthemum

14th Anniversary ~ Orchid

15th Anniversary ~ Red Rose

20th Anniversary ~ Aster

25th Anniversary ~ Iris

30th anniversary ~ Lily

40th Anniversary ~ Gladioli

50th Anniversary ~ Yellow roses and violets

Other Floral Connotations Worth Mention

Amaryllis ~ pride

Anemone ~ I am forsaken

Aster ~ patience

Azalea ~ take care of yourself

Bachelor button ~ single blessedness

Bells~of~Ireland ~ good luck

Bittersweet ~ truth

Bluebell ~ humility

Calendula (marigold) ~ cruelty, grief, jealousy, valor

Calla ~ beauty

Carnation (pink) ~ I'll never forget you

Carnation (purple) ~ capriciousness (not sure)

Carnation (red) ~ my heart aches for you, admiration

Carnation (solid) ~ yes

Carnation (yellow) ~ disdain

Chrysanthemum ~ truth, friendship

Clover ~ providence

Coreopsis ~ always cheerful

Crocus ~ cheerfulness

Cyclamen ~ resignation and goodbye

Daffodil ~ regard, unequaled love, new beginnings

Daisy ~ innocence, loyal love, purity

Dandelion ~ faithfulness, happiness

Daylily ~ coquetry

Daylily (Tiger lily) ~ wealth, pride

Delphinium ~ joy

Fern ~ secret bond of love

Forget me not ~ true love, memories

Forsythia ~ anticipation

Gardenia ~ secret love

Geranium ~ stupidity, folly

Gladiolus ~ strong character, Give me a break, I'm really sincere, you pierce my heart

Gloxinia ~ love at first sight

Grass ~ submission

Hellebore ~ anxiety

Hibiscus ~ delicate beauty

Holly ~ defense, domestic happiness

Hollyhock ~ fertility

Hyacinth ~ rashness, play

Hydrangea ~ thank you for understanding, heartlessness, and frigidity

Iris ~ faith and hope

Ivy ~ wedded love, fidelity, friendship, affection

Jonquil ~ desire, love me

Larkspur ~ playful, frivolous

Lilies ~ life, chastity, innocence, purity

Lily of the Valley ~ sweetness, humility, return to happiness

Magnolia ~ nobility

Marigold (calendula) ~ cruelty, grief, jealousy

Mistletoe ~ kiss me, affection

Monkshood ~ beware

Moss ~ maternal love, charity

Narcissus ~ egotism, formality

Nasturtium ~ conquest

Orchid ~ beauty and love

Peony ~ shame, happy life and marriage

Petunia ~ resentment, anger, your presence soothes me

Poppy ~ eternal sleep, oblivion, imagination

Primrose ~ I can't live without you

Rose (in full bloom) ~ I love you

Rose (burgundy) ~ unconscious beauty

Rose (orange) ~ fascination

Rose (peach) ~ modesty, gratitude, appreciation, admiration, sympathy

Rose (pink) ~ grace and admiration

Rose (purple) ~ enchantment

Rose (red and white) ~ Unity
Rose (red) ~ love, respect, courage
Rose (white) ~ purity, secrecy
Rose (yellow) ~ joy, gladness, freedom, jealousy, infidelity
Rosebud ~ youth and beauty
Snapdragon ~ deception, gracious lady
Stephanotis ~ happiness in marriage
Stock ~ bonds of affection, promptness
Sunflower ~ haughtiness
Sweet Pea ~ goodbye, blissful pleasure,
 thank you for lovely time
Tulip ~ perfect lover, fame
Tulip (red) ~ believe me, declaration of love
Tulip (variegated) ~ beautiful eyes
Tulip (yellow) ~ sunshine and smiles
Violet ~ faithfulness and modesty
Wisteria ~ will you dance with me?
Zinnia ~ lasting affection, constancy, goodness

Bibliography

The Meaning of Flowers, University of Illinois – US Dept of Agriculture – Local Extension, January 2017

http://web.extension.illinois.edu/cfiv/homeowners/080131.html

Seaton, Beverly. *The language of flowers: a history*. University of Virginia Press, 2012.

Dumont, Henrietta. *The Floral Offering: A Token of Affection and Esteem; Comprising the Language and Poetry of Flowers*. HC Peck & Theo. Bliss, 1852.

Phillips, Henry. *Floral Emblems*. Saunders and Otley, 1825.

Shoberl, Frederic, ed. *The Language of Flowers: With Illustrative Poetry; to which are Now Added the Calendar of Flowers and the Dial of Flowers*. Lea & Blanchard, 1848.

The Language of Flowers, Brochure from Iowa State University Extension

Heilmeyer, Marina. The Language of Flowers: Symbols and Myths. 2001.

Kirkby, Mandy. A Victorian Flower Dictionary. 2011

Durant, Mary. Who Named the Daisy? Who Names the Rose? 1976.

Elliott, Brent. *The Victorian Language of Flowers*. 2015.

Zeman, Anne M. *Fifty easy old-fashioned flowers*. Macmillan, 1995.

Trudier Harris, "'The Yellow Rose of Texas': A Different Cultural View" in Juneteenth Texas: Essays in African-American Folklore, ed. Francis Edward Abernethy et. al. (Denton, Texas: University of North Texas Press, 1996).

Harrison, Lorraine. "RHS Latin for gardeners." *United Kingdom: Mitchell Beazley* (2012)

Crowell, Robert L. *The Lore & Legends of Flowers*. Crowell, 1982.

Scoble, Gretchen, and Ann Field. The Meaning of Flowers: Myth, Language & Lore. Chronicle Books, 1998.

Boland, Maureen, and Bridget Boland. Old wives' lore for gardeners. Farrar, Straus and Giroux, 1977.

Adachi, M., C. L. E. Rohde, and A. D. Kendle. "Effects of floral and foliage displays on human emotions." *HortTechnology*10.1 (2000): 59-63.

MacDougall, Elisabeth Blair. Ars hortulorum: sixteenth century garden iconography and literary theory in Italy. 1972.

Armitage, Allan M. Armitage's manual of annuals, biennials, and half-hardy perennials. Timber Press, 2001.

Bunting, Eve. *Sunflower house*. Houghton Mifflin Harcourt, 1996.

Dobelis, Inge N. "Magic and medicine of plants." *Pleasantville, NY: Reader's Digest Association* (1986)

Martin, Laura C. *Wildflower folklore*. The Globe Pequot Press, 1993.

Martin, Laura C. *The folklore of trees and shrubs*. Globe Pequot Press, 1992.

Allen, David Elliston. "Victorian fern craze." (1969).

Drew, John K. Pictorial guide to hardy perennials. Merchants Publishing Co, 1984.

Thompson, Ken, 'The Amazing Secret of the Scarlet Pimpernel', The Telegraph June 2015

http://www.telegraph.co.uk/gardening/plants/11686207/The-amazing-secret-of-the-scarlet-pimpernel.html

Laufer, Geraldine Adamich. Tussie-mussies. Workman Pub., 1993.

Gips, Kathleen M. *Flora's dictionary: the Victorian language of herbs and flowers*. Village Herb Shop, Incorporated, 1995.

Tussie-Mussies, Laufer GA. "The Victorian Art of Expressing yourself in the Language of Flowers." (1993).

Skolnick, Solomon M., and Lana Kleinschmidt. *The Language of Flowers*. Peter Pauper Press, 1995.

Flowers & Their Meanings, http://www.flowermeaning.com/

Bullen, Annie. *Language of Flowers*. Jarrold Publishing, 2004.

Mercatante, Anthony S. *The Magic Garden: The Myth and Folklore of Flowers, Plants, Trees, and Herbs*. New York: Harper & Row, 1976.

Engelhardt, Molly. "The Language of Flowers in the Victorian Knowledge Age." *Victoriographies* 3.2 (2013): 136-160.

About the Author

Olivia Wylie is a professional landscaper who specializes in the restoration of neglected gardens. When the weather keeps her indoors, she enjoys researching and writing about the plants she loves and the ways they've shaped human thought. She lives in Colorado with a very patient husband and a rather impatient cat.

www.ingramcontent.com/pod-product-compliance
Lightning Source LLC
Chambersburg PA
CBHW040255100426
42811CB00011B/1275